Space Scientist

COMETS
AND METEORS

Heather Couper

Franklin Watts

London   New York   Toronto   Sydney

© 1985 Franklin Watts
Limited

First published in 1985 by
Franklin Watts Limited
12a Golden Square
London W1R 4BA

First published in the USA
by Franklin Watts Inc.
387 Park Avenue South
New York, N.Y. 10016

First published in Australia
by Franklin Watts
Australia
1 Campbell Street
Artarmon, NSW 2064

UK ISBN: 0 86313 267 7
US ISBN: 0-531-10000-6
Library of Congress
Catalog Card No: 84-52570

Illustrations by
Drawing Attention
Rob Burns
Eagle Artists
Ron Jobson
Michael Roffe

Photographs supplied by
NASA
Space Frontiers Ltd

Designed by David
Jefferis

523.6
C

Printed in Italy

Space Scientist

COMETS
AND METEORS

Contents

Cosmic debris

Although we often talk of space as being empty, this certainly isn't true of the region of space which surrounds the Earth. Think about all the space junk up there for a start: old satellites, discarded rocket stages and bits of protective casing, to name just a few examples. But there is natural debris in space, too. Our Solar System – the family of nine planets and one star to which the Earth belongs – is littered with rubble. And that's what this book is all about.

Cosmic rubble isn't all that obvious until it springs into action. And that's when we get to see it. Nearly everybody has watched a

"shooting star" streak across the sky before disappearing in a flash. In fact, shooting stars – or meteors, to give them their proper name – are among the least substantial of the cosmic rubble, although they sometimes look very impressive. Meteors are tiny clumps of powdery dust which are attracted to the Earth by gravity, and then plunge headlong into the atmosphere. Because they travel so fast, they burn up in a flash – rather like a re-entering spacecraft without a heat shield!

Not all meteors are quite so tiny, and some really large and solid ones almost survive the drop through our atmosphere. These fireballs can look like huge firework rockets, trailing sparks behind them as fragments of the meteor break off under the strain. The biggest meteors of all do succeed in landing and, once down, are called meteorites.

▽ Trailing sparks, a gigantic fireball plunges into the Earth's atmosphere, disintegrating as it goes. Only the biggest meteors of all can survive the fiery descent, and these may end up forming craters many kilometers across.

Because more than two-thirds of the Earth is covered by seas, that's where most meteorites fall – and are consequently lost for good. But every year, a number of meteorites do land on solid ground, allowing astronomers to get their hands on something which has spent all its life in space. The biggest meteorites of all land with such enormous energy that they vaporize completely on impact. Although nothing of the meteorite remains, the blast can make a huge crater.

But for most people, the most exciting bits of cosmic debris are the comets. Ghostly and mysterious, a bright comet can hang nearly motionless in the sky for days, looking like a dagger about to strike. In reality, the comet is way out in space and traveling very fast. And despite its dramatic appearance, it's just a ball of ice being heated relentlessly by the Sun. Its evaporating gases are forced out behind into a tail by the pressure of the Sun's solar wind. But there's nothing which quite compares with the appearance of a comet in the sky. And we're due for a visit by one very soon.

△ Although a great comet can be the most spectacular sight in the sky, its appearance is all show. A comet is little more than a ball of ice which is fiercely heated when it approaches the Sun. The escaping gases can form a tail which stretches for millions of kilometers.

Where did they come from?

Comets, meteors and meteorites are all debris left over from the formation of our Solar System 5,000 million years ago. It was then that a huge cloud of dust and gas – a nebula – began to collapse under its own pull of gravity. Most of the gas collapsed right to the center, where it became extremely compressed. As a result, the central ball of gas grew very hot. Eventually, it turned on its own nuclear energy source, and the shrinking stopped. The gas ball had become a star – our Sun.

All around the young Sun sticky dust grains swirled and collided. Some stuck together, building up bigger and bigger clumps which were to form the cores of the young planets. But not all the chunks were used up. Some continued to hit the planets after they had been born, forming huge craters. Others, lying close to Jupiter, were unable to knit themselves into a single body because of the strong pull of Jupiter's gravity. Fragments of rock from this asteroid belt continue to hit Earth as meteors or meteorites today.

△ The Orion Nebula, where radiation from young stars is making the surrounding gas glow. Nebulas are places where new stars and planets are forming even today.

Far beyond the planets – stretching perhaps a quarter of the way to the next star – lies the domain of the comets. Astronomers think that millions of frozen comet nuclei, like giant icebergs, live there in a huge, spherical cloud which surrounds the Solar System. The cloud is like a tidemark of the old nebula from which we formed. But once a comet leaves the cloud, it can never return. Forced to circle the Sun forever, it eventually wears away to dust.

◁ The Solar System as it looked 4,500 million years ago, a swirling mass of dust and debris.

▽ The Solar System as it is today. Debris still remains in the asteroid belt and comet cloud.

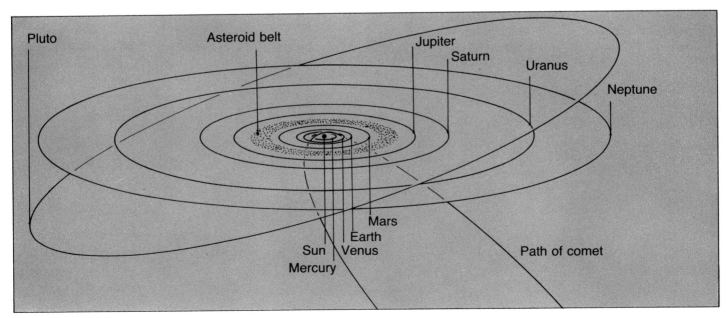

Pluto
Asteroid belt
Jupiter
Saturn
Uranus
Neptune
Mars
Earth
Sun
Venus
Mercury
Path of comet

Comets today

The sight of a brilliant comet, with its tail stretching halfway across the sky, is something that people remember all their lives. You may meet people today who were lucky enough to see both Halley's Comet and the Daylight Comet back in 1910; for many of them, it's as if it was yesterday. But we haven't been so lucky in the second half of the twentieth century. There hasn't been a really big, nearby comet for ages. But comets are unpredictable. It could be tomorrow when one falls out of the comet cloud and plunges in toward the Sun – and it could be a big one. The bigger the comet, the more ice there is to be evaporated by the Sun, and the more spectacular the comet will appear.

One way to become immortal is by discovering a comet for the first time. If you're the only discoverer, then the comet is named after you alone – Comet West, or Comet Austin, perhaps. But quite often, several people around the world discover a comet simultaneously. Then the comet is named after the first three discoverers, like Comet Tago-Sato-Kosaka. Why do so many comets have Japanese names?

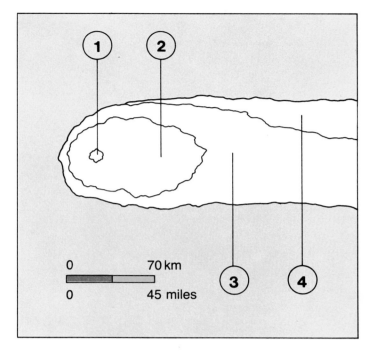

△ Probes like Giotto will tell us how a comet is really made up. At present, we have only a rough idea. At the comet's center, there may be a small rocky core (1); a coma of evaporating gases (2); and then gas and dust tails (3 and 4).

It's because Japanese amateur astronomers make a special hobby out of comet-hunting. They buy the biggest binoculars they can afford – the kind that need a permanent stand – and they sweep the dawn and sunset horizons for fuzzy patches every morning and night. This way, they stand the chance of catching a comet when it's on its way toward the Sun, but still too far off to be very active. But it's not easy – and it gets very chilly. You need an expert knowledge of the sky, and – something that many comet-hunters have invested in – a pair of electrically heated slippers!

▽ The elongated path of Halley's Comet around the Sun takes 76 years to complete. Its orbit isn't circular, like the (almost) circular paths of most of the planets, because the comet first plunged in from a long distance away. In 1986, the European Giotto probe (right) will intercept the comet.

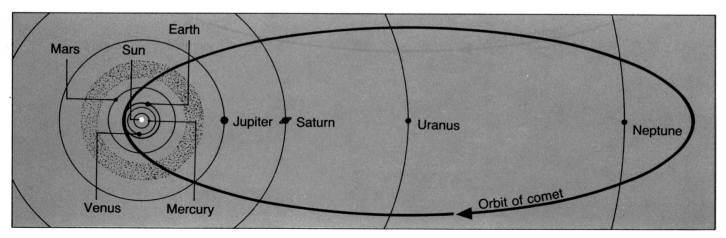

The life and death of a comet

Most comets never put on any kind of show at all. They spend their entire lives millions of kilometers away from the Sun in the huge cloud of comets which astronomers think surrounds our Solar System. But from time to time, the almost imperceptible pull of a remote passing star can dislodge a comet from its frozen perch. Perhaps ten times each year, a "new" comet plunges in toward the Sun.

A few comets will actually collide with the Sun itself – a fate which doesn't affect the Sun at all, but is certain death for the comet. The majority have enough "sideways" momentum to swing around the Sun, although a comet is very rarely able to get back to where it came from.

Once a comet has taken the first plunge, it is usually trapped among the planets forever by the strong pull of the Sun and of the giant planet Jupiter. Comets like this which return to the Sun regularly, are called short-period comets. Long-period comets, on the other hand, have enough initial momentum to carry them far away from the planets' pulls, and may not return for millions of years.

Nearly all comets have very elongated orbits which bring them in close to the Sun, and then out a long way again. This is the result of the comet's long plunge, and it means that a comet behaves very differently depending on whether it is close to the Sun or far away. No one knows what a comet looks like when it is at aphelion (at its furthest from the Sun). It's likely to be a frozen ball of ices – including methane ice, ammonia ice and water ice – a few kilometers across. It's not certain whether a comet has a small core. But comets all have

▽ Sequence showing the development of Halley's Comet's gas tail in 1910, when it last came close to the Sun. The tail grew in length as the comet approached, and shrank again as the comet receded. The tail's appearance changed from day to day; often it contained jets and knots of gas. The Earth actually passed through the tail.

◁ No one knows what a comet looks like in close-up. But it is certain that a comet's ices sublime, or evaporate without melting first.

△ Halley's Comet on its last visit in 1910 was photographed at the Yerkes Observatory. Background stars appear as trails.

some rock in the form of fine dust particles, although these may amount to no more than a dirty coating on the icy surface.

When a comet comes within the orbit of Mars (or sometimes Jupiter), the Sun's heat starts to evaporate some of its icy surface. A huge halo of gas known as the coma begins to envelop the frozen nucleus. Some of the biggest comas recorded have been even larger than the Sun, which is 1.4 million km (864,000 miles) across – although comet comas are very rarefied.

As the comet approaches the Sun, it speeds up, accelerated by the strength of the Sun's pull. The gases boil away more and more fiercely, and are forced away from the comet's head by the solar wind, a gale of charged particles streaming from the Sun. As a result, the comet develops a long straight tail of gas, which always points away from the Sun. It drops dust around its orbit too, which forms a fainter, gently curved dust tail.

When a comet rounds the Sun, it is at its most spectacular. Its tails may stretch across the sky for millions of kilometers. But its moment of glory is brief – perhaps a week or two at most. Soon it is heading back into the depths of the Solar System, tail-first, and freezing as it goes. With every return more and more of the comet evaporates – until only a swarm of dust remains.

Some famous comets

Comets can be notoriously unpredictable. They can brighten or fade without warning, or be deflected from their orbits by the influence of a planet (usually Jupiter). Or they can disappear altogether! Comet Biela, discovered in 1772, did just that. It returned regularly every six years until 1846, when it suddenly split into two. The "twins" returned on time in 1852, although the distance between them had increased to 2 million km (1.25 million miles). Since then, neither has been seen – although showers of meteors (almost certainly comet dust in this case) have been observed when the comet should have appeared.

▽ Comet West was a magnificent sight in the dawn skies of early 1976, particularly from the southern United States. But then it amazed everybody by splitting into four – probably because the Sun's pull broke up its nucleus.

Another famous "disappearing comet" was Comet West. This long-period comet came to us out of the blue in late 1975, when it was discovered on a photographic plate by Richard West of the European Southern Observatory in Chile.

As the comet approached the Sun, some people predicted that it would become extremely bright. Yet even as it rounded the Sun, the comet remained obstinately faint, although it was just visible to the unaided eye in clear skies.

But on emerging from the Sun's glare, it surprised everyone by suddenly brightening some 100 times! It was so brilliant that in some parts of the world it could be seen during daylight hours – and its tail appeared to fan into five separate tails. But its moment of glory was shortlived. In March 1976 observers noticed that the comet's nucleus was beginning to break up. First two fragments were seen, and later four. Then one disappeared altogether. Of the other three, two may never return again. Only one will possibly come back – in 500,000 years time.

The most famous comet of recent times was first discovered by a satellite – the orbiting Infra-Red Astronomical Satellite (IRAS) – in April 1983. While news of confirmation of the discovery by Earth telescopes was being delayed by a public holiday in the UK, two amateur astronomers spotted it almost simultaneously. The comet was named IRAS-Araki-Alcock in their (and the satellite's) honor. It was George Alcock's fifth comet discovery.

Comet IRAS passed closer to the Earth – 5 million km (3.1 million miles) – than any other comet bar one; and many people remember seeing it in the sky in early May 1983. It will not return again for many centuries.

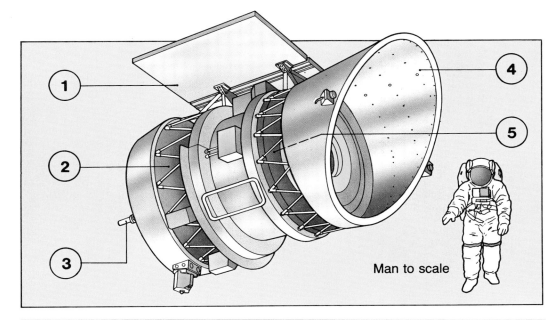

Man to scale

◁ The IRAS satellite detected lukewarm radiation from bodies in space. It found six new comets and many other new objects.
1 Solar panel
2 Infrared-detecting mirror
3 Radio antenna
4 Sunshade
5 Additional heat-detecting experiments (inside)

▽ Color-coded IRAS picture of Comet IRAS-Araki-Alcock, showing the warmest parts of the comet as yellow (center) fading to blue (coolest).

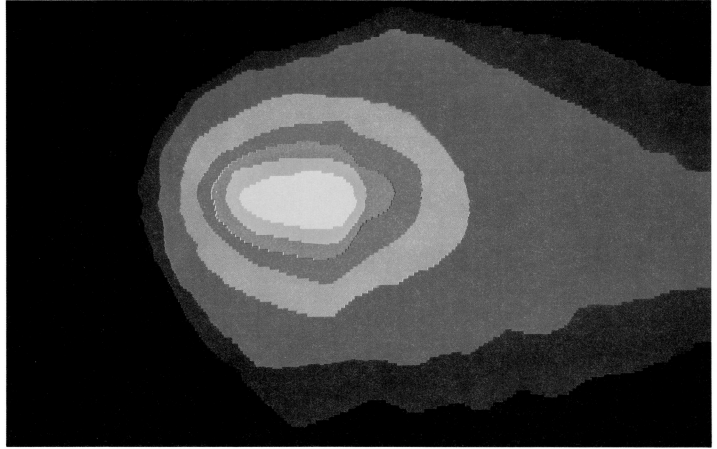

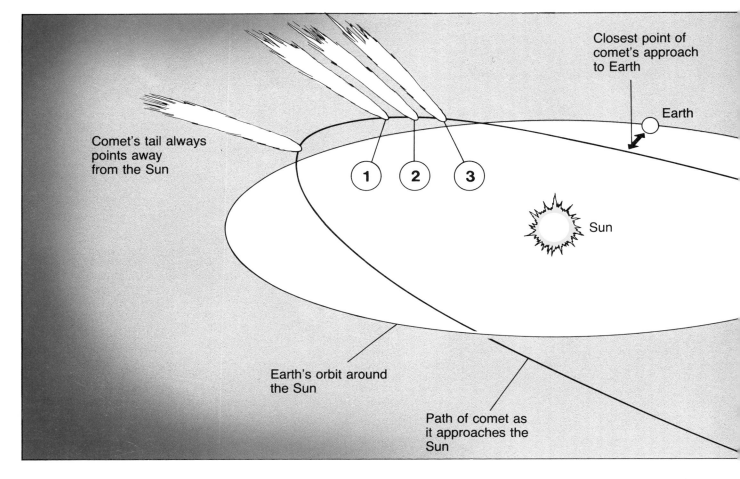

Closest point of comet's approach to Earth

Earth

Comet's tail always points away from the Sun

Sun

Earth's orbit around the Sun

Path of comet as it approaches the Sun

Mission to Halley's Comet

△ The Giotto probe is named after the Italian artist Giotto di Bondone, who used the 1301 appearance of Halley's Comet as the model for his star in *The Adoration of the Magi*.

Of all the comets, the most famous by far is, of course, Halley's Comet. Edmond Halley didn't actually *discover* it, but it was he who noticed that comets had regularly appeared in 1531, 1607 and 1682. He suggested that this could in fact be one and the same comet and predicted its return. Right on schedule the comet reappeared in 1758 (after his death), and so was duly named after him.

Halley's Comet has been returning to the inner Solar System roughly every 76 years since perhaps 240 BC. Its last visit was in 1910, when it put on a fine show in spring skies (although it was rather overshadowed by the even brighter Daylight Comet). Many were actually very worried about the comet colliding with the Earth; and some even took "Comet Pills" to prevent being poisoned by the gases in its tail!

Halley's Comet is right on course for its next visit to us in 1985–6. It is already visible in very big telescopes, and it's getting brighter by the moment as the Sun's heat does its work. But this will be one of Halley's less spectacular visits, because the comet will not

14

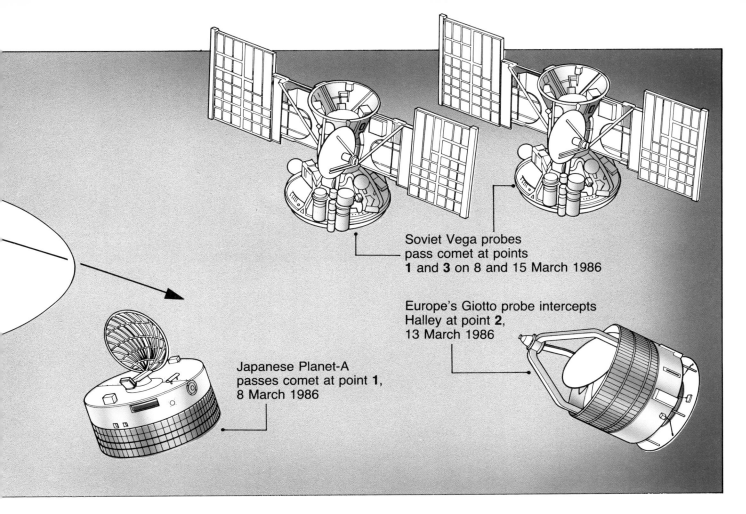

Soviet Vega probes
pass comet at points
1 and **3** on 8 and 15 March 1986

Europe's Giotto probe intercepts
Halley at point **2**,
13 March 1986

Japanese Planet-A
passes comet at point **1**,
8 March 1986

come at all close to the Earth. People in the southern hemisphere should get a good view in April 1986, but northern hemisphere dwellers will be less lucky.

The best views of all, however, will be reserved for the four spaceprobes due to rendezvous with the comet in March 1986. There are two Soviet craft called Vega; a Japanese probe called Planet-A; and a European probe called Giotto. There would have been an American probe, too, but it didn't receive Government funding.

The two Vega probes will be flying to the comet via the planet Venus, where they will drop landing probes. On arriving at the comet, one will probably fly by at a safe distance, while the other will aim right for its head. Both probes carry cameras and instruments to analyze the composition and the outflow rates of gases boiling off Comet Halley.

The small Japanese probe Planet-A will devote itself to studying the growth and shrinking of the huge hydrogen gas coma around the comet. It will also analyze how the solar wind interacts with the comet's gases.

△ Positions of the Earth, Halley's Comet and the four spaceprobes at the time of the comet's next visit to the inner Solar System in 1985–6.

The probes will intercept the comet after it has rounded the Sun and is at its most active. We can expect some spectacular pictures.

Planet-A will not approach closer than 150,000 km (93,000 miles).

It's Giotto, however, which will have the most exciting mission. After launch by an Ariane rocket, the probe will spend eight months in space before plunging deep into the comet's coma. Scientists hope to fly Giotto within 500 km (310 miles) of the nucleus itself. On board will be 10 different experiments, including cameras, telescopes, and instruments to monitor conditions in the comet's neighborhood. But it is also a suicide mission. Giotto will encounter the comet at a speed of 68 km/s (42 mps) – that's nearly 250,000 km/h (155,000 mph). At such speeds, collisions with even tiny particles of dust in the comet's nucleus will be fatal. Although Giotto has a dust shield, nothing will save it when it reaches the heart of Halley's Comet.

Meteor showers

Once a comet becomes trapped in the inner Solar System, the only way it can go is downhill. On each encounter with the searing heat of the Sun and the force of the solar wind, the comet wears away just that little bit more. After 100,000 years – a short period in astronomical timescales – very little of the comet remains except for grains of dust.

Even while the comet is still "alive," it litters its orbit with dust – like a wheelbarrow trailing soil. If the Earth happens to cross the orbit of the comet during its own journey

around the Sun, the dust grains sweep into our atmosphere. Traveling at speeds of up to 50 km/s (30 mps) the tiny particles burn up through friction with the air at a height of about 100 km (60 miles) above the ground. We see this as a brief streak of light in the sky – a shooting star, or meteor. For a second or so after the meteor has passed, you may see its trail slowly fade in the sky, as the excited air around its track cools down. And then it's all over – until the next piece of comet dust sweeps in.

Because we plow into so many dust grains when we cross the orbits of dead or dying comets, we get to see a lot of meteors at once at certain times of the year. On these occasions, we have meteor showers. The duration of the shower, and the number of meteors we see, depends on how quickly we cross the parent comet's orbit, and how dense the dust is. If the comet itself is anywhere near the Earth at the time, the extra dust associated with it will guarantee even more meteors.

On every orbit about the Sun, the Earth crosses the paths of a number of comets. This leads to a succession of about two dozen meteor showers, which occur at the same times each year. The showers are named after the star pattern occupying the point in the sky from which the meteors appear. For instance, the plentiful meteors which occur in early

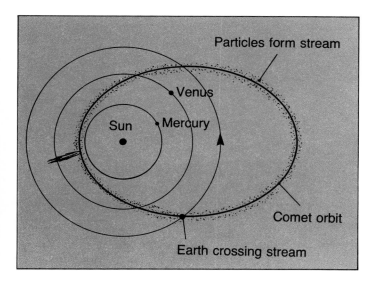

△ Shower meteors are grains of dust produced as comets gradually wear out. As the comet ages, the particles spread themselves into a stream all around the comet's orbit. If the Earth crosses the stream, we see a meteor shower.

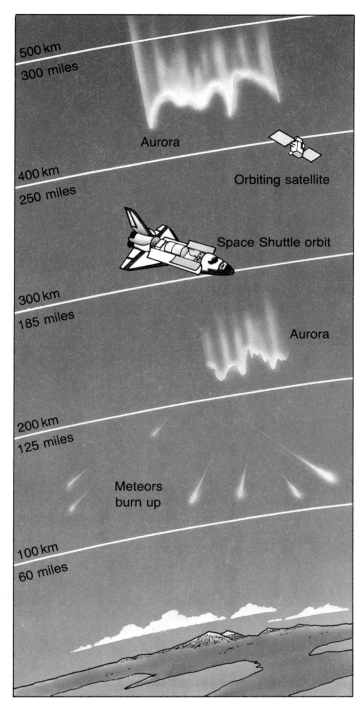

500 km
300 miles

Aurora

400 km
250 miles

Orbiting satellite

Space Shuttle orbit

300 km
185 miles

Aurora

200 km
125 miles

Meteors
burn up

100 km
60 miles

◁ The Earth's atmosphere acts like a shield to protect us from particles and radiation coming in from space. But some particles can penetrate. The aurora, for instance, is a glow caused by energetic charged particles from the Sun which hit the atmosphere. Fast-moving meteors penetrate even deeper.

▽ This old woodcut gives an impression of the Leonids meteor storm which took place in 1833. The storm lasted for only a few hours, and it occurred because the parent comet, Comet Tempel-Tuttle 1866 I, was close to the Earth at the time. Most meteor showers are not nearly as spectacular as this.

August seem to come from the direction of the constellation Perseus – and so they're called the Perseids. And although the meteors plunge into our atmosphere in parallel lines, the effect of perspective makes them appear to diverge from a point in the sky instead. This point, called the radiant, allows you to work out whether the meteor you've just seen is a shower meteor or not. If you can trace back its path to the radiant, then it's one from the shower; if not, it's a random "sporadic" meteor. There's a list of meteor showers on pages 26–7, as well as some tips on meteor watching.

How many meteors can you expect to see? On a really clear night, a dense shower (like the Perseids in August, or the Geminids in December) should produce over 50 meteors per hour – but remember, you do need an unobstructed view! Most showers are much less spectacular. But you can never tell. In 1966 the normally modest Leonids shower produced a storm of 100,000 per hour over the central USA, although for a brief period only.

The asteroid belt

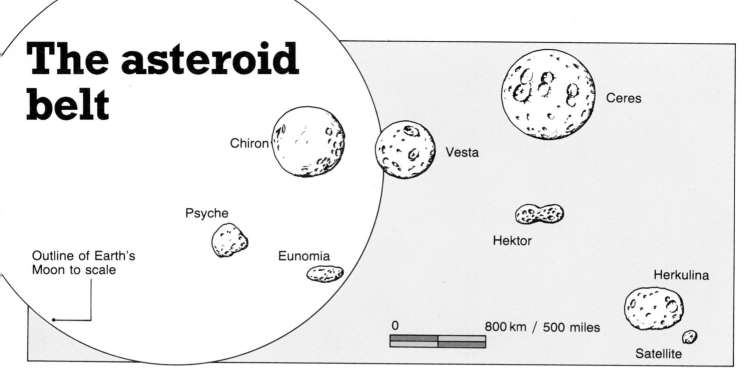

Chiron

Vesta

Ceres

Psyche

Eunomia

Hektor

Herkulina

Outline of Earth's
Moon to scale

0   800 km / 500 miles

Satellite

By no means all the meteors we see come from dying comets. Many appear from any direction, at any time – in fact, even on a night when there isn't a meteor shower, you should be able to spot more than six an hour. These sporadic meteors are often more substantial than shower meteors, so they penetrate deeper into the atmosphere and look brighter. A few appear even brighter than the brightest planets, and are called fireballs. The sight of a fireball trailing sparks as it blazes across the sky is one you never forget.

These more substantial meteors are mainly broken-off bits of asteroids – a swarm of shattered rocky fragments which circle the Sun between the orbits of Mars and Jupiter. Some asteroids have orbits outside the main belt and can travel into the inner Solar System, or as far out as Saturn. One of the most extreme of these is Icarus, 1.4 km ($\frac{7}{8}$ mile) in diameter, which passes within 30 million km (19 million miles) of the Sun at its closest in, but retreats to a distance of 300 million km (186 million miles) at its furthest. Asteroids like these – which must be alternatively baked and frozen – can give rise to meteors if they shed fragments when they cross the Earth's orbit.

Other extreme asteroids include Chiron, 600 km (375 miles) in diameter, which was discovered in 1977 and orbits the Sun between Saturn and Uranus; and mysterious Earth-crossing asteroid 1983 TB, found by the IRAS satellite, which could even be the core of a dead comet. Some astronomers think that the tiny planet Pluto – only 2,400 km (1,500 miles) across – could be an asteroid itself, the biggest of an undiscovered swarm at the edge of our Solar System.

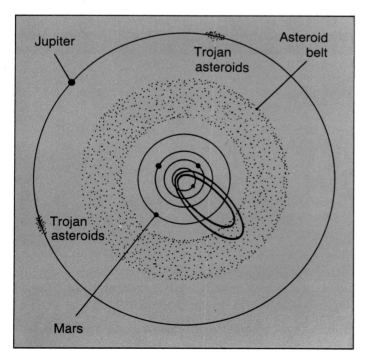

Jupiter

Trojan asteroids

Asteroid belt

Trojan asteroids

Mars

◁ The asteroids occupy a zone between the orbits of Mars and Jupiter where people once predicted that there ought to be a planet. There are probably over 100,000 chunks of rock in the belt, but together, they would make a body smaller than the Moon.

The vast majority of asteroids belong firmly in the asteroid belt. First to be discovered was Ceres, 1,000 km (620 miles) in diameter, in 1801, after a special search for an apparently "missing" planet between Mars and Jupiter had been set up. Ceres – visible in binoculars – turned out to be by far the largest of the asteroids, but there must be thousands which are only the size of boulders. Because the asteroids are so small, it's difficult to get much information about them. Only now are astronomers beginning to understand how asteroids are made, and what they're composed of. The latest research seems to indicate that some asteroids have little moons of their own in orbit about themselves!

The asteroids are most certainly not fragments of an "exploded planet," as is sometimes claimed. They're probably bits of a

△ The peculiar almost-double asteroid Hektor seems to be shaped rather like a peanut! In this imaginary picture set in the future, a probe surveys the system, and assesses its potential for mining. One day, we may obtain many of our construction materials from the asteroids.

very small world which lay too close to the giant planet Jupiter ever to knit together into one. If Jupiter hadn't been around, the planet formed would hardly have been substantial – about one-tenth as massive as our Moon!

The asteroid belt has not proved the hazard to space travel that people once predicted. Four probes have traveled through the belt without accident, and it's certain that more will follow. One of these may investigate whether it would be possible to mine the asteroids for their valuable minerals in the future.

Meteorites

A few really solid meteors of the "asteroid-fragment" variety manage to survive their fiery descent through the Earth's atmosphere without burning away altogether. When they land on the Earth's surface, they become meteorites. But despite the fact that something like 100,000 tons of meteoroidal debris rains down upon the Earth each year, meteorites are surprisingly difficult to come across. It's partly because two-thirds of our planet is covered in water, and so most meteorites land in the oceans. Those which do fall on land look at first sight like ordinary Earth rocks. But it's important for scientists to be able to recognize them, because they provide the only hard evidence – apart from Moon rocks – of what bodies from space are really like.

Many people expect meteorites to be hot when they land, but because their outer casing is all burned away by then, they are surprisingly cool. And it's not true that meteorites inflict widespread injury. The only definite meteorite casualty was a dog, sadly killed when a meteorite landed in Egypt.

◁ This meteorite, discovered in 1982 in Antarctica, has a composition very similar to Moon rock, and may have come from the lunar highlands.

When a meteorite falls through the atmosphere, it usually breaks up into dozens of fragments before landing because of the enormous stresses it undergoes. Some meteorites are more fragile than others. The most fragile are stony meteorites, which also erode most quickly when they reach the Earth's surface. Although iron meteorites are much less common – the ratio of stones to iron is about 10:1 – they're much stronger, and more easily recognized on the ground. Stony-iron meteorites come somewhere in-between.

The real value of meteorites to astronomers, though, is that they're made of material which has remained virtually unaltered since our Solar System was born. They're a key to our past.

◁ A team of explorers stands on the biggest meteorite in the world, which they discovered in 1920. The Hoba West iron meteorite lies near Grootfontein in Namibia, southwest Africa, but no one saw it fall. It weighs over 60 tons and it has never been moved. You can see smaller examples of meteorites in museums all over the world.

METEORITE CHART

Stony meteorites

Chondrites

These are by far the most common kind of meteorite. They are made of fragments of stone, together with small round bodies called chondrules. Each chondrule is about a millimeter across and made of a silicate material. Some very rare chondrites – the carbonaceous chondrites – contain carbon, the element on which life is based.

Achondrites

Achondrites are also stony meteorites, but they are much less common than chondrites. They have no chondrules, and are generally rather fine-grained. Many achondrites are similar to basalt rocks on the Earth, and they may well be fragments of lava flows covering the bodies which later broke up to form the meteorites.

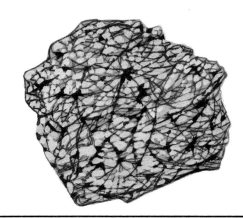

Iron meteorites

Iron meteorites can be polished to highlight their beautiful crystalline internal structure. They are made of a mixture of nickel-iron, together with smaller amounts of copper, cobalt, phosphorus and iron sulfide. Once on Earth, they survive longer than other kinds of meteorite. They were once part of the cores of their parent bodies.

Stony-iron meteorites

Of the three basic types of meteorite, stony-irons – or siderolites – are the rarest of all. They contain almost equal quantities of nickel-iron metal and stony minerals, including such minerals as olivine, pyroxene and plagioclase. They seem to have cooled extremely slowly and were probably formed deep inside their parent bodies.

Meteorite craters

Today, most meteorites land on the Earth (or on any planet for that matter) without forming a crater. Most of them are just too small. But in the past, not long after our Solar System was formed, there were many much larger meteoroidal fragments around. These were able to make very big craters. The huge craters we see on the Moon – and also on Mercury, Mars and the moons of Jupiter and Saturn – nearly all date back to this violent episode of bombardment long ago, when the larger bodies in the Solar System "mopped up" the smaller ones.

There are meteorite craters on the Earth, too, but they aren't nearly so obvious. Most of them have been heavily eroded by the effects of wind, rain and frost. They are also more sparsely distributed because, in the past, many meteorites must have fallen into the sea.

Our atmosphere also keeps the numbers down by burning even very large meteors as they enter. But the main reason why meteorite craters aren't as common on the Earth as on some other planets is because of the Earth itself.

The Earth is still an active planet. Heat currents rising from deep inside "drive" around the platforms or plates upon which the light rocks of the continents float. This process of plate tectonics means that the land surface of the Earth is never still. When continents collide, mountain ranges are formed; when they separate, the gap is filled with new crust welling up from below. And so Earth's surface is constantly changing. You only find craters on the ancient, little-changed "shield" areas which make up parts of continents like Canada and Australia. The crust elsewhere in the world is too new to bear any scars.

Unlike volcanic craters, which are conical and steep-sided, meteorite craters are saucer-shaped, sometimes with a central peak. Their shape is the result of an explosion – meteorite craters are not just "holes in the ground."

When a meteorite lands, it is abruptly stopped, and its enormous energy is converted into heat and shock waves. As the meteorite burrows underground, shock waves shatter and compress the rocks above. But when the pressure is released, there is a huge explosion. The area around the newly opened crater is showered with debris. If the meteorite is a really massive one, it won't even be able to survive the impact, but will be completely vaporized by the process instead.

△ **Left** Thousands of enormous meteorites bombarded the Moon's surface in the past, creating craters which are sometimes many hundreds of kilometers across. Some of the "splashes" from these impacts probably hit Earth later on, and so became lunar meteorites.

△ The Moon's crater-scarred surface looks much as it did thousands of millions of years ago, because the Moon has no atmosphere to cause erosion, and no plate tectonics. Its dark plains are craters which are so deep that lava from inside the Moon later welled up to flood them.

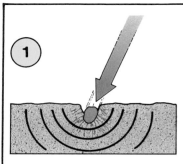

1 Meteorite hits ground. Shock waves compress and shatter the underlying rocks.

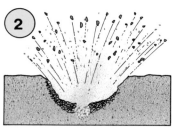

2 Shattered rocks rebound from the impact and there is a huge bomb-like explosion.

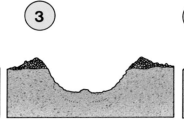

3 After the explosion a gaping hole is left. Debris builds up the "shoulders" of the crater.

4 The finished crater. There may be a central peak created as a result of the rebound.

Some famous meteorites

About 65 million years ago a huge meteorite 20 km (12½ miles) across and a million million tons in weight smashed into the Earth. It blasted a crater 200 km (125 miles) in diameter and threw up a pall of fine, pulverized dust which enveloped our planet for many years. When it dispersed, virtually nothing remained alive. Starved of sunlight, all the plants had died. The animals which fed on them followed soon after; and finally, the carnivorous animals died too. Most important of all, the dinosaurs – masters of the Earth for the previous 135 million years – were completely wiped out.

Science fiction? Many scientists think not. It's certain that the dinosaurs did vanish rapidly, as did many other creatures at about the same time. Geologists analyzing sediments deposited some 65 million years ago find an abrupt break in the number of fossils laid down around that time – many before, but very few after. Occupying the "break" is a thin layer of iridium – a substance rare on Earth, but common in meteorites. Could this iridium, which is deposited worldwide, have come from the meteorite and its dust pall? And if so, where is the crater the meteorite left behind?

Although it's possible that the meteorite fell in the sea, leaving a hidden scar, not everyone agrees that this catastrophe actually happened. Far more certain is the meteorite impact which took place near Flagstaff,

△ The Arizona Meteorite Crater was made about 25,000 years ago by the impact of a body 70 m (230 ft) across and 250,000 tons in weight.

▷ No one is absolutely sure just what the Siberian Fireball was, but its blast had the energy of an H-bomb.

Arizona, about 25,000 years ago. Today the scar is there for all to see in the form of the Arizona Meteorite Crater. The crater measures 1.2 km (¾ mile) across, 180 m (590 ft) deep, with a raised rim 45 m (145 ft) high. Most of the 250,000-ton meteorite was vaporized by the impact, although a little survived to litter the surrounding area with iron fragments.

Have any really big meteorites hit the Earth recently? One possible contender is the giant Siberian Fireball. Early in the morning of 30 June 1908 a brilliant object streaked across the skies of Siberia, trailing smoke and giving off loud bangs. The driver of the Trans-Siberian express thought his train had exploded! It landed – or appeared to land – in a remote and desolate area, which explorers didn't succeed in reaching until 1927. There they found – nothing! There were flattened trees and reindeer skeletons over thousands of kilometers, but no meteorite, and no crater. Astronomers now think that the fireball was the head of a small comet which exploded in the atmosphere. But there are still people around who believe it was an alien starship!

▽ The end of the road for the dinosaurs: some astronomers believe that they were wiped out by the impact of a meteorite the size of a small asteroid.

Meteor watching

You can see meteors on any night of the year. But your best chance to see them in large numbers is during a meteor shower, when the Earth plows into clouds of comet debris during its orbit about the Sun. There's a list of the best meteor showers at the bottom of this page.

You don't need a telescope or even binoculars for meteor watching. In fact, it's best just to use your eyes, because then you get the widest view. You will need a star map to be able to identify the area of sky the meteors come from, and you'll need a comfortable chair and a sleeping bag. Remember, you could be staring at the sky for hours on end!

Always write down details of what you see. List the times of all the meteors (useful to have a friend with an accurate watch making notes for you!), along with their direction and trail length (plot on a star map), and how bright each one is compared to the stars.

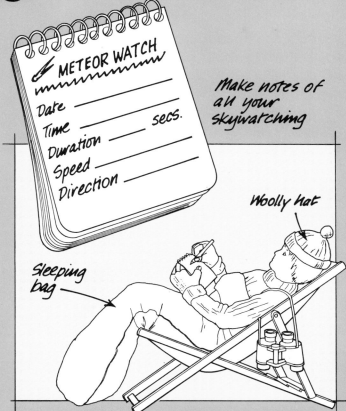

Make notes of all your skywatching

Woolly hat

Sleeping bag

Meteor shower dates

Name of Shower	Maximum activity	Maximum number an hour	
Quadrantids	3–4 January	50	
Lyrids	22 April	10	
eta Aquarids	5 May	10	The numbers given here are the most that you are ever likely to see in an hour. Expect fewer and you won't be disappointed!
delta Aquarids	31 July	25	
Perseids	12 August	50	
Orionids	21 October	20	
Taurids	8 November	10	
Leonids	17 November	10	
Geminids	14 December	50	

Taking photographs

Meteor photography is not easy! Meteors move fast, so you can "catch" only the brightest on film. A fast film is important for a start. So is a camera with a B (brief) setting. This allows you to keep the shutter open for as long as you want, so you get a better chance of capturing a meteor. A steady support is essential!

For best results, point your camera half-way "up" the sky, and two hand-spans away from the radiant. This way you'll see the most meteors. Now open the aperture to its fullest to let in the most light (f 2.8 setting). Finally open the shutter for 15–20 minutes. If you watch the same area of sky, you'll be able to time any meteors which come in. And hopefully, one or two will be on your film!

B-setting on camera

Time exposures require a "B" setting

Use a mini-tripod to get a steady image

You can try to get pictures like this

Observing Halley's Comet

Although this will be one of Halley's Comet's less spectacular visits, you should easily be able to see it away from city lights. If you miss it, you'll have to wait for its return in 2062! The chart below shows how the comet will move against the star background over the 1985–6 period.

In November 1985 people in the northern hemisphere should first start to see it with binoculars. During December and January it will brighten to a dim naked-eye level and grow two short tails. In February it will be behind the Sun. It will emerge, much brighter, in March – but it will be steadily moving south. April will be Halley's month of glory – for southern observers. Its tail may stretch over a quarter of the sky. After this the comet will start to fade, although moving north again. After May, it will be invisible once more.

Path of the comet

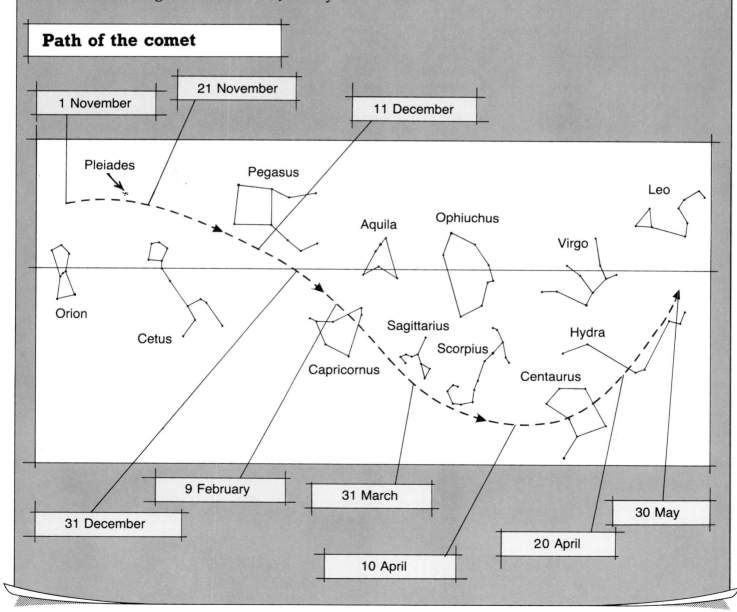

Halley facts

Halley's Comet was first seen in (at least) 240 BC, and has been recorded on every return since, except 163 BC.

Halley's Comet was seen before the Battle of Hastings in 1066. People believed it was a portent of England's defeat by France.

In May 1910 the Earth passed through the tail of Halley's Comet.

Debris from Halley's Comet is believed to give rise to two meteor showers – the Orionids and the eta Aquarids.

The comet is estimated to have a mass of 50 million million million tons, and its gas tail stretches for 32 million km (20 million miles) at its closest approach to the Sun.

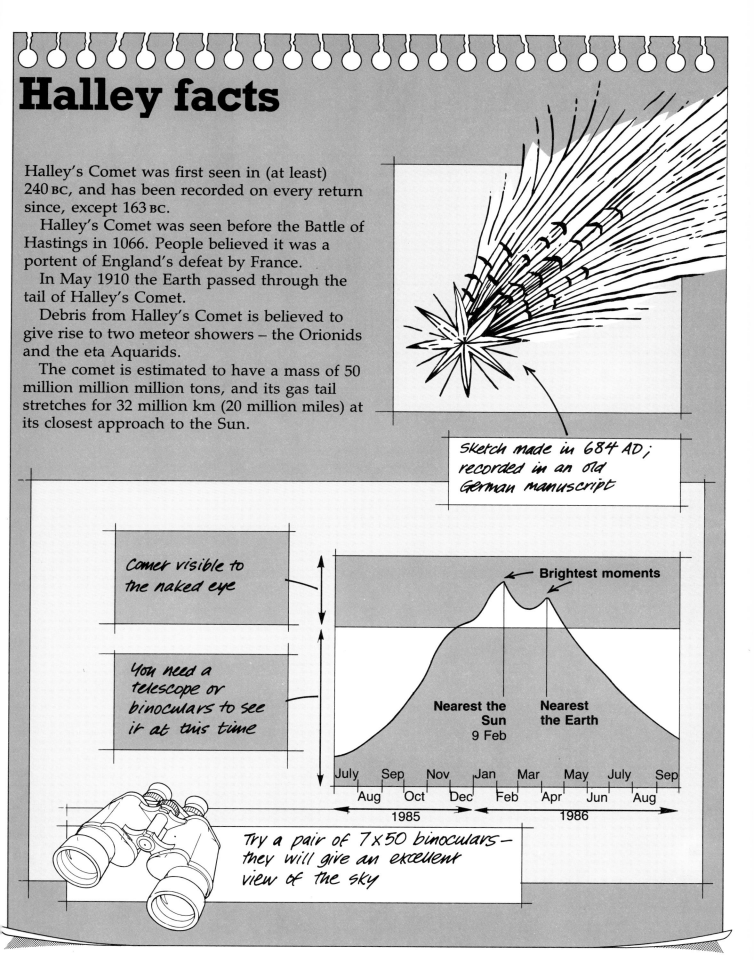

Sketch made in 684 AD; recorded in an old German manuscript

Comet visible to the naked eye

You need a telescope or binoculars to see it at this time

Brightest moments

Nearest the Sun 9 Feb

Nearest the Earth

July Sep Nov Jan Mar May July Sep
Aug Oct Dec Feb Apr Jun Aug
1985 1986

Try a pair of 7 x 50 binoculars – they will give an excellent view of the sky

Glossary

Here is a list of a few words you might come across in connection with comets, meteors and meteorites. Many of them refer to more advanced ideas we haven't had space to cover in this book – but at least it means you won't get stumped by an unfamiliar word.

Achondrite Stony meteorite made of fine-grained rock without tiny, silicate drops called chondrules.

Aphelion Point in a body's orbit when it is farthest from the Sun.

Apollo group Group of asteroids with orbits from the main belt into the inner Solar System.

Asteroid One of many thousands of minor planets.

Carbonaceous chondrite Rare type of chondritic meteorite (see below) containing life-like carbon inclusions.

Chondrite Stony meteorite containing 1 mm silicate drops called chondrules.

Coma The huge halo of gas which forms the head of a comet. It consists of evaporated ice from the nucleus or core of the comet.

Comet Dusty, icy body which can develop a huge gaseous coma and long tail when it approaches the Sun.

Constellation An apparent grouping of stars in the sky, usually named after a mythological figure.

Crater Saucer-shaped hole blasted out by an explosion.

Dust Grains of "space-soot" blown off the surface of very cool stars.

Eccentric Having an elongated orbit, like a comet.

Ellipse An oval-shaped orbit. Most bodies move around the Sun in ellipses.

Fireball A meteor brighter than any of the planets.

Giotto European probe due to intercept Halley's Comet.

Gravity Force of attraction between bodies.

Mass The amount of matter that a body contains.

Meteor Chunk of rock in space which burns up when it enters the atmosphere.

Meteorite Chunk of rock or metal large enough to survive the journey through the atmosphere and land.

Meteoroid Small chunk of rock or metal out in space.

Meteor shower Swarm of meteors seen when the Earth crosses the orbit of a comet.

Moon A planet's satellite.

Nebula Cloud of gas and dust in space, part of which may collapse to form stars and planets.

Nucleus (of comet) The frozen core of a comet. It may have a solid center.

Oort Cloud Reservoir of frozen comets thought to surround the Solar System in a huge, spherical cloud.

Orbit Path traveled by a body in space.

Path The track of a meteor across the sky.

Perihelion Point in a body's orbit when it is closest to the Sun.

Planet Large body which orbits the Sun, or another star, in its own right.

Planet-A Japanese probe due to intercept Halley's Comet.

◁ A portent of doom: Halley's Comet hangs in the sky as Harold prepares for the Battle of Hastings in 1066, where the French defeated him. (From the Bayeux Tapestry.)

Places to visit

There are plenty of observatories to visit on the West Coast. And if you're near Flagstaff, Arizona, don't miss a trip to Meteor Crater! But one of the best ways to get really involved in astronomy and space is to join your national or local astronomical society.

Radiant Point in the sky from which shower meteors appear to come.
Retrograde Having an orbit which travels "backward" – clockwise around the Sun as seen from above, instead of counterclockwise. Many comets have retrograde orbits.
Siderite Metal meteorite composed mainly of nickel-iron.
Siderolite A stony-iron meteorite.
Solar wind Stream of energetic charged particles blowing away from the Sun. It forces comet tails to point away from the Sun's direction.
Sporadic meteor Meteor not associated with a meteor shower.
Star Massive body which is able to generate its own light and heat through nuclear fusion reactions.
Tail (of comet) Outflow of gas which sublimes off a comet's nucleus as the comet approaches the Sun. Comets also drop dust grains around their orbits, generating a broader, fainter dust tail.

△ Tiny Enceladus, one of the moons of Saturn, is covered in meteorite craters which have walls made of ice!

Tektite Small glassy body about 1 cm (¼ in) across, associated with meteorite craters on Earth. It may be splashed-up debris from the impact forced briefly into space, which melted as it re-entered the Earth's atmosphere.
Train The luminous trail of heated (ionized) air left behind after a meteor has passed.
Trojan group Family of asteroids which move around the Sun in the same orbit as Jupiter.
Vega Pair of Russian probes which are due to intercept Halley's Comet via Venus.
Zenith The point in the sky directly overhead.
Zenithal Hourly Rate (ZHR) The number of shower meteors per hour that an ideal observer would see if the radiant were overhead. Rates are always lower than this.

The leading society for people who are really serious about observing meteors is the American Meteor Society, Dept. of Physics & Astronomy, SUNY, Geneseo, N.Y. 14454. And if you want to help with the observations of Halley's Comet on its next visit, drop a line to the International Halley Watch, c/o Stephen Edberg, JPL, 4800 Oak Grove Dr, T-1166, Pasadena, Calif. 91109. Beginners should join up with their local astronomy club instead – that way, you can have a lot of fun, and do things like going on organized meteor watches. Astroclubs often take their members to visit places of astronomical interest. Among these are planetariums – artificial star theaters where you can watch the sky in warmth and comfort. You can even see artificial meteor showers, too!

The US and Canada have some of the best planetariums in the world. There isn't space to give a detailed list here, but you'll find some of the biggest in the following cities: Chicago, Los Angeles, New York, Salt Lake City, Toronto, Tucson, Vancouver.

Many local museums have collections of meteorites, and observatories often sell big posters of comets and meteors as well as books and magazines.

△ In 1966 the parent comet of the Leonids meteor stream, Comet Tempel-Tuttle 1866 I, came close to the Earth and caused a meteor storm. For 15–20 minutes, observers in the USA saw more than 2,000 meteors a minute!

31

Index